ISBN: 979-8-218-65383-5
Printed in The United States of America
First Edition

ACKNOWLEDGMENTS

Thanks to Cateran Society broadsword instructors Shane Clune, Eóghan Shoger, and Matt Park for demonstrating the techniques shown in this book. Thanks also to A Heart's Desire Photography and the members of the Wolves of Dunvegan, who assisted with a few of the older photographs, and to Phread Cichowski, for the use of genuine Highland broadswords from his collection in a few of the older pictures.

Introduction 6
Basic Exercises
Guards
Cutting
Targe Exercise
The Penicuik Guards

The Highland Charge Exercises............20

The Ten Lessons of Sword and Targe 26
Lesson One
Lesson Two
Lesson Three
Lesson Four
Lesson Five (Basic)
Lesson Five (Advanced)
Lesson Six
Lesson Seven (Advanced)
Lesson Eight
Lesson Nine
Lesson Ten

The Bouting Exercises 44
Highland Singlestick Game
The Leg Attack Game
The Binding Game
The Lifting Game
The Escaping Game
The Single Combat Game
The Pitching Game
The Hanging Thrust Game
The Highland Charge Game
The Countering Game

The Lessons of Thomas Page 46

1: The Slip
2: The Feint
3: Field Attacks
4: Disabling
5: Pitching to a Hanging
6: The Leg Attack
7: The Bind
8: The Lift
9: Escaping the Bind
10: Escaping the Lift

The Penicuik Lessons 50

1: The Hanging Guard
3: The Bind
4: The Lift
5: The Escape
6: Single Combat
7: Pitching to a Hanging
8: The Hanging Thrust
9: The Highland Charge
10: Countering the Charge

Loose Play 52

Advanced Sword and Targe Lessons 54

Plain Playing
Timing
Slipping the Leg
 Slipping the Body
Double Attacks
The Feint
The Invitation
Actions on the Blade
Commands
Counter-Disarms
Sword, Targe and Dirk Lessons
Certification in Sword and Targe

Introduction

In August 1608, a formidable expedition was organized under Lord Ochiltree for the purpose of securing the long-deferred submission of the islanders. The terms to which the chiefs were to be called upon to submit were strictly specified by the Privy Council, and they included... the abstaining from the use of guns, bows, and two-handed swords, and the confining themselves to single-handed swords and targes.

(Mitchell, Dugald. 1900. A Popular History of the Highlands and Gaelic Scotland From the Earliest Times Till the Close of the 'Forty-Five. Paisley: Alexander Gardner, pp 450-451)

With this expedition to the Hebrides – intended to force the clan chiefs to submit to the authority of the central government – the Privy Council of Scotland put an end to the era of the two-handed sword in the Scottish Highlands. For the next 138 years, Highland warriors were known for their expertise with sword and targe, a weapon combination they used throughout the civil wars of the 1640s, the 1688 Battle of Killiecrankie, and the uprisings of 1715 and 1745, as well as many clan battles and lesser skirmishes in between.

This was also the era of the Highland Charge, a battlefield tactic in which the Highlanders would try to drive the enemy from the field of battle with a single ferocious rush. The Highlanders would fire their muskets or pistols, draw their swords, shout out their battle cries, and run straight at the enemy. If the enemy army panicked and ran – as was often the case – the Highlanders would simply cut them down from behind.

As simple as this approach may seem, it should really be seen as a tactically intelligent adaptation to the realities of early firearms. If the enemy army fired a volley of musket shot, the Highlanders could cross the battlefield at a dead run and be on them before they could reload for a second volley. Faced with the prospect of being cut down by the Highland broadsword, the enemy soldiers would often break and run.

The methods of using the sword and targe were not completely preserved, but there is some discussion of their use in a handful of old fencing manuals. Donald McBane, a famous Highland soldier, discusses the targe in passing in his Expert Sword-Man's Companion of 1728. Thomas Page, an English broadsword fencer who sold Highland broadswords in his shop, gives a selection of sword and targe techniques in his The Use of the Broadsword from 1746. This work has been somewhat controversial in the historical fencing community, as not everyone is convinced that Page's sword and targe techniques were genuine. My personal view is that Page had been exposed to at least a handful of real techniques with the sword and targe, even if he had never been taught a complete system. The "Penicuik Sketches," also from 1746, show a number of sword and targe stances as interpreted by an eyewitness to the 1745 Jacobite uprising. Finally, Alexandre Valville's Treatise on the Contre-Pointe from 1817 shows an interesting illustration of the sword and targe fencing Valville had presumably seen during his time in Scotland.

The method taught in this book is a combination of all these sources, based on a foundation of historical 18th century broadsword play as taught in multiple different fencing manuals. These manuals include the works of Donald McBane (1728), Thomas Page (1746), Andrew Lonnergan (1771), John Ferdinand (1788), G. Sinclair (1790), Archibald MacGregor (1791), Henry Angelo (1799), John Taylor (1804), and Thomas Mathewson (1805).

Garde du montaguard Écossois. | Oudyaucamentos uaeuvaie copueno Maxuundaqo.

Valville's illustration of "The Guard of the Highlander," a variation on the hanging guard

The Bind, a technique from Page's Use of the Broadsword

The Lift, a technique from Page's Use of the Broadsword

Basic Exercises

Before you can practice with a training partner, you'll need to learn a set of basic skills. If you're already an experienced broadsword fencer, there won't be much here you haven't seen before – but pay special attention to exercises 8-10. If you've never studied broadsword before, you'll need to practice all ten of these exercises regularly for at least six months before you attempt to bout with the sword and targe.

There are two stances you'll need to learn before you begin. These are sometimes known as "narrow" and "wide," and in the Cateran Society they are also known as second position and fourth position; however, none of these terms are used in the broadsword manuals and they are used here only for convenience.

Narrow: begin by standing with your hands on your hips and your feet together, facing forward. Turn your rear foot to a 90-degree angle and lower your weight a little. Step forward with your lead foot on bent knees so that your feet are in line with each other and about one-and-one-half foot-lengths apart. Keep most of the weight on the rear leg. This is the "narrow" or "second position" stance. In the narrow stance, both feet are said to be on the "line of defense," an imaginary line running between you and your opponent. The narrow or second position stance is the same as the "on guard" position in modern fencing.

Wide: step back with your lead foot so that your feet are almost parallel, with the lead foot just slightly ahead of the rear foot, and the feet about shoulder-width apart, so you are standing almost square to your imaginary opponent. Turn your rear foot to a forty-five-degree angle as you do so. This is the "wide" or fourth position stance. This stance is used to traverse and was also used with some guards. In the wide stance, the line of defense runs between your feet.

Advancing and Retreating

Advancing and retreating footwork is similar to modern fencing footwork, and you won't use it often with the sword and targe. However, that's not to say you won't use it at all, and it's a good place to start. Advancing and retreating is done in the narrow stance.

This is how Thomas Page defines the advance and retreat:

To Advance.

Is to press upon your Adversary under the Cover of some Guard, Step by Step, with the right Foot always before; making but half Steps at a Time.

Retreating.

Is retiring from him under the Cover of some Guard by half Steps, the left Leg moving first Backwards, and the right drawing after it. (Page, 1746)

Page goes on to describe the stance to be used in the advance and retreat before describing both actions in more detail:

The Position of the Body must be very erect, its Center of Gravity kept exactly over the Left Leg, with the Right Foot a little advanc'd, that the whole Weight of the Body may rest over the Left Foot, and the Right be at absolute Liberty for Motion. From this Posture the Steps to be learn'd are as follow: the Advance, the Retreat, and the Traverse.

The Advance

When the erect Attitude above describ'd is obtain'd both for Grace and Use, step forward with the Right Foot about one third of your Lunge, and at the same Time transfer so much of the Weight of your Body form your Left Leg on to your Right, as may enable you to slip your Left Foot along the Ground, (not lifting it off) up towards your Right Heel, and stopping within half a Foot thereof; at which Moment step forward again with the Right Foot, and alternately repeating the same Steps advance as far as is necessary, still preserving an erect firm and graceful Attitude through every Motion of the Advance.

The Advantage of this Step is gaining Space in the length of Ground, and pressing so upon your Adversary, as to oblige him to retreat from you unto worse Ground, or some disadvantageous Situation.

The Retreat

From the same erect Attitude before describ'd, transfer the Weight of your Body almost wholly from the Left to the Right Leg, so that you may be fully enabled to step backward with your Left Foot, lifting it clear from the Ground, the better to avoid any unevenesses that cannot be seen behind, and setting it firmly down about sixteen Inches backward, draw back the Right Foot within twelve Inches of the Left, but not lifting it off the Ground; and repeating these Steps also alternately, retreat as far back as you find useful.

Above: The narrow stance

Below: The wide stance

The Advantages of this Step is by retiring either to dray your Adversary from the advantageous Ground he is in Possession of, or to gain a more advantageous Ground that lies behind you; or to avoid the Difficulties into which you are fallen, by your Adversary's pressing too closely upon you, and engaging you with superior Strength up to half Sword, and very often all these three Advantages are obtained at the same time. (Page, 1746)

Note that Page's advance and retreat are a little different than the advance and retreat used in modern fencing, in that the following foot drags the ground and is not lifted up. In the advance, the rear foot is dragged, and in the retreat, it is the front foot that is dragged. This is to make your footwork stable on uneven ground.

To practice advancing and retreating, pick a target such as the opposite wall of your training area, and advance until you reach it before retreating to your starting point.

Traversing

The traverse is used much more often in sword and targe than the advance and retreat. Here is Page's description:

To Traverse.

Is stepping from the straight Line either to the Right of Left in a Circle, still preserving the Center of that Circle, in the Center of the Line...

This also begins from the same erectness and firmness of Posture, and is twofold viz. The Fore Traverse, and the Back Traverse. The Fore Traverse is performed in a large Circle, the Center of which is the Middle of the Line of Defence, on which Line you and your Adversary stand; such is the Line P. Q. C. H. G. in the opposite Page, and the Circle form'd by the Traverse will be, P. A. C. E. G. I. L. N. For the Right Foot being at Q. and the Left at P. the traverse is begun by stepping about with the Left Foot from P. to A. and the Right Foot immediately after from Q. to B. and then the Line A. B C. K. I. will be the Line of Defence; at the next Step, remove the Left Foot from A. to C. and then the Right from B. to D. which will make the Line C. D. C. M. L. the Line of Defence; and you wil be still faceing C. the Center of that Circle, which you are now Traversing, an the Middle of every Line of Defence; proceed also in the same Manner with the Left Foot from C to E and the Right Foot from D to F. then will E F. C. O. N. be the Line of Defence; in the same Manner proceed

to G. H; to I. K; to L. M; to N. O; and to P. Q; which is the Place from which you set out, and you will have successively the Lines G. H. C. Q. P.; I. K C. B. A; L. M. C. D. C; N. O. C F. E; for Lines of Defence; and now you are come about to the Line P. Q. C. H. G; which was the Line of Defence when you began to Traverse.

The Back Traverse

Is the counter Part of the Fore-Traverse, doing every Step backwards as in that is done forwards; as for Example, standing in the Line of Defence P. Q. C. H. G. with the Right Foot at Q. and the Left at P. begin the Back Traverse with removing the Right Foot from Q. to P. and the Left from P. to N. both in the Line N. O. C. F. E. which will then be the Line of Defence; and then by removing the Right Foot from O. to M. and the Left from N. to L. you have L. M. C. D. C. for the Line of Defence; and in the same Manner going backwards through K I, H G, F E, D C, B A, you will arrive at Q P, from whence you began the Back Traverse.

The Advantages of these two Traverses are very great, as will be explained more at large in the Action of Fighting; but their Advantages in gaining Ground may be known here: If in the Retreat you are stopt behind by a Wall, Ditch, or any other Impediment, you may by beginning either Traverse which ever you find most convenient to extricate your self with ease, and gain Ground either to the Right or Left; and if you Traverse half the Circle, it will bring your Adversary into the very same Difficulties from which you departed. (Page, 1746)

As confusing as this may be, it isn't really as difficult as it might seem. To perform the traverse, step from the narrow stance into the wide stance by moving your rear foot to the left. Then step from wide stance into narrow stance by moving your lead foot to the left. This is the fore traverse: it will cause you to circle the opponent in a clockwise direction.

To perform the back traverse, step from narrow to wide by moving your lead foot to the right, then step from wide to narrow by moving your rear foot to the right. The back traverse will cause you to circle the opponent in a counterclockwise direction. (These directions should be reversed if you are left-handed.)

Whenever you're traversing, think of the opponent as being in the center of a large circle. This is the circle shown in Page's footwork diagram.

The fore traverse will cause you to rapidly close in on your opponent, and the back traverse will take you away from your opponent. If you want to prevent yourself from closing in on the opponent in the fore traverse, step back a little every time you take a step with the rear foot.

To practice the traverse, take a narrow stance and make the fore traverse until you come back around to your starting point. Then make the back traverse until you come back to your starting point in the other direction.

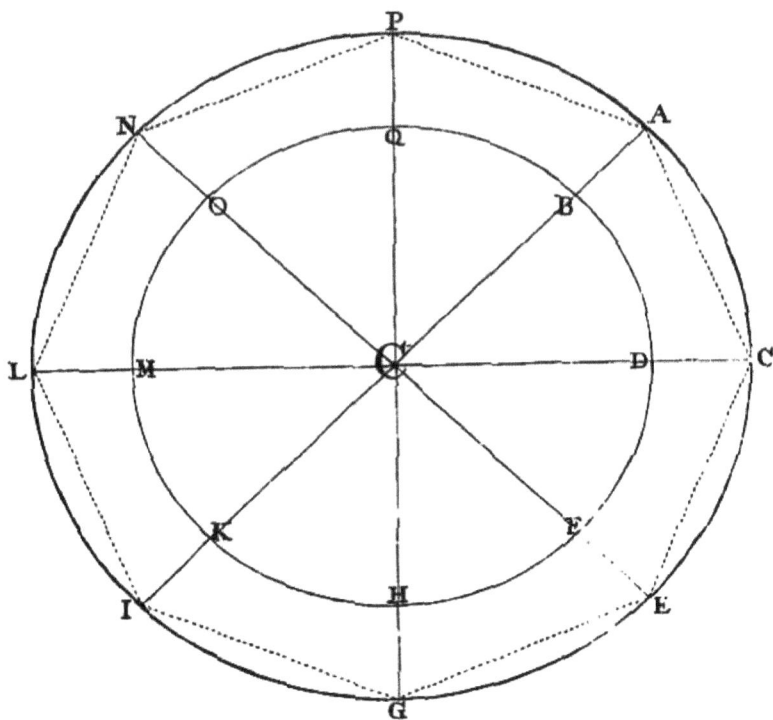

Page's footwork diagram. With your feet at Q and P you are narrow. With your feet at Q and A you are wide. The opponent is at C

Passing footwork is also used with the sword and targe. To practice passing, all you have to do is walk forward with one foot passing the other, as in an ordinary walking step. Now do the same thing with deeply bent knees, keeping your weight low and your stance relatively wide. That's passing. Passing footwork in sword and targe is often used on a full run, while charging at the enemy.

Guards

There are three basic guards with the sword and targe. Do not hold the targe flat in front of your body in any of these guards. As McBane says:

This target is of great use to those who rightly understand it. But to unexperienced people is often very fatal, by blinding themselves with it, for want of rightly understanding it. Therefore who has a mind to use it must take care to have it upon an edge so as to cover his left side, from which is a defence against ball or any weapon. (McBane, 1728)

Having the targe "upon an edge" means to hold it in such a way that the rim of your targe is aimed at your opponent's left shoulder, leaving some space in which you can see despite being behind the targe.

Outside Guard: stand in the wide position, with your targe hand slightly further forward than your sword hand. Hold your sword so that your point is at the level of your opponent's right eye, with your wrist turned outward, basket angled to the right and fingernails down. Your sword arm should be bent rather than completely extended, and your sword blade should cross the body diagonally from the right to the left. The Penicuik Sketches show a variation I call the "high outside guard." To take the high outside guard, simply raise your sword from the outside guard until it is high above your head, ready to make an outside cut without needing to chamber the blow. Some of Page's sword and targe techniques also work more easily from a high outside guard.

Passing forward and attacking

Passing forward and attacking

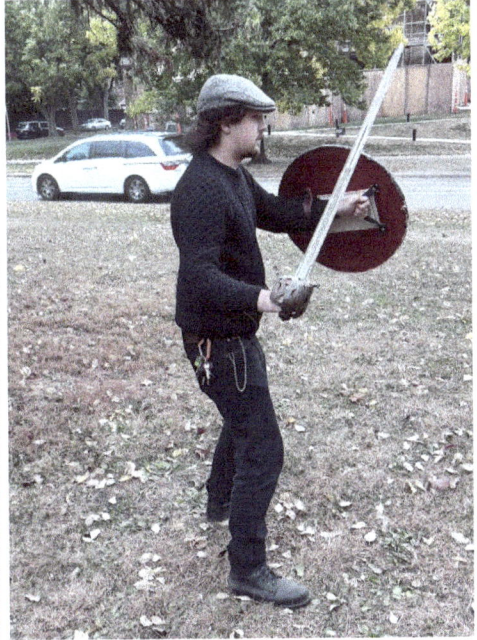

Above (Left): The outside guard from the front. (Right): Side view

Below: The high outside guard

Hanging Guard: to take the hanging guard, extend your sword arm, with your hilt high over your head and to the outside a little. The edge should be upward, as if aimed at the sky. The blade should slope diagonally downward to cross the body, and your point should be aimed at your opponent's body. There is more than one version of the hanging guard. The version described here is a "seconde" hanging guard. If you bend your elbow in and let the blade hang down to your left from this position, you will be in a "prime" hanging guard. The hanging guard can be used in either a wide or narrow stance.

A seconde hanging guard

A prime hanging guard

Above: The hanging guard in the narrow stance (front)
Below: Above: The hanging guard in the narrow stance (side)

Inside Guard: to take the inside guard, step into a narrow stance and rotate your wrist so that your sword point is aimed at your opponent's left eye. Now your edge should be angled to the left with fingernails up. Your sword blade should cross the body diagonally from the left to the right. When I take the inside guard with sword and targe, I prefer to tuck the basket-hilt of the sword up against the inside of the targe.

To practice the guards, take an outside guard followed by a high outside guard, a prime hanging guard, a seconde hanging guard, and finally an inside guard. Repeat this sequence ten times:

Outside

High Outside

Prime Hanging

Seconde Hanging

Inside

The inside guard (front)

The inside guard (side)

The inside guard (side)

Cutting

There are seven cuts in the later broadsword system, but only five are used in sword and targe. The later manuals number the seven cuts, but the earlier manuals use only names rather than numbers. For a right-handed fencer, cut 1 is a diagonal, descending cut from the right. Cut 2 is a diagonal, descending cut from the left. Cut 3 is a diagonal, ascending cut from the right. Cut 4 is a diagonal, ascending cut from the left. Cuts 5 and 6, the horizontal cuts, are not used with sword and targe. Cut 7, straight down at the opponent, is rarely used with sword and targe. Cut 1 is called an "inside." Cut 2 is called an "outside." Cut 3 is called a "low inside." Cut 4 is a "low outside." Cut 7 is a "medium."

To practice cutting, stand in an outside guard and cut inside, outside, low inside, and low outside in sequence, or 1,2,3,4. Repeat the exercise ten times. Occasionally add a medium cut.

The inside cut (1)

The outside cut (2)

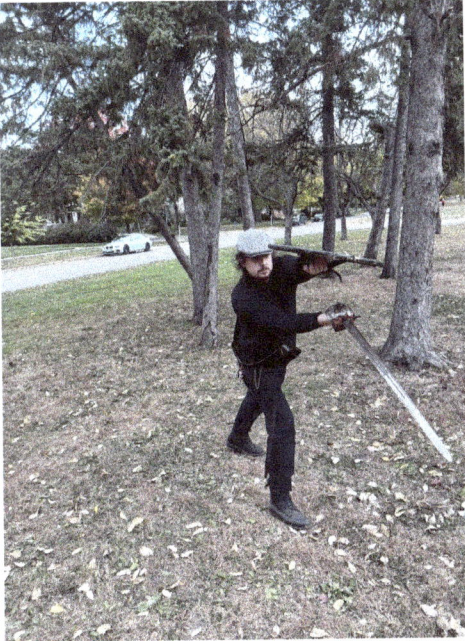

The low inside cut (3)

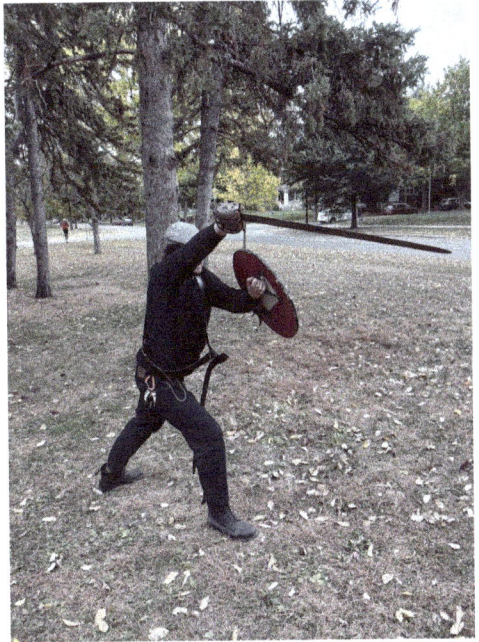

The low outside cut (4)

The medium cut (7)

Figure Eight 1 and 2

One of the best ways to practice cutting is to perform figure eights. If you practice hundreds of repetitions of these figure eights every chance you get, you will develop the skill to cut effectively with a sharp sword. You will also learn how to fence for a long time without getting tired, as your body will teach itself the most efficient way to move as you perform the repetitions.

To perform figure eight 1 and 2 with sword and targe, take an outside guard in the wide stance. Cut inside as you step narrow, then outside as you step wide again, and continue cutting as you traverse. Always step narrow on your inside cut and wide on your outside cut. Try to work up to one hundred cuts, and then to one thousand cuts if possible.

Upper figure-eight: cut 1

Upper figure-eight: cut 2

Figure Eight 3 and 4

To perform figure eight 3 and 4 with sword and targe, take an outside guard in the wide stance. Cut low inside as you step narrow, then low outside as you step wide again, and continue cutting as you traverse. Always step narrow on your inside cut and wide on your outside cut.

Lower figure-eight: cut 3

Lower figure-eight: cut 4

Suppression

Suppression is my term for using the targe to suppress any counterattack as you attack your opponent. To practice suppression, take a high outside guard and practice stepping in with a chop to the outside or a low inside cut. (In other words, make either cut 2 or cut 3). Just before you cut with your sword, punch your targe out while turning your targe to a horizontal position to suppress any counterattack. The idea is not to make hard contact with the antagonist's sword, but simply to be in position to immediately suppress any attempted counter.

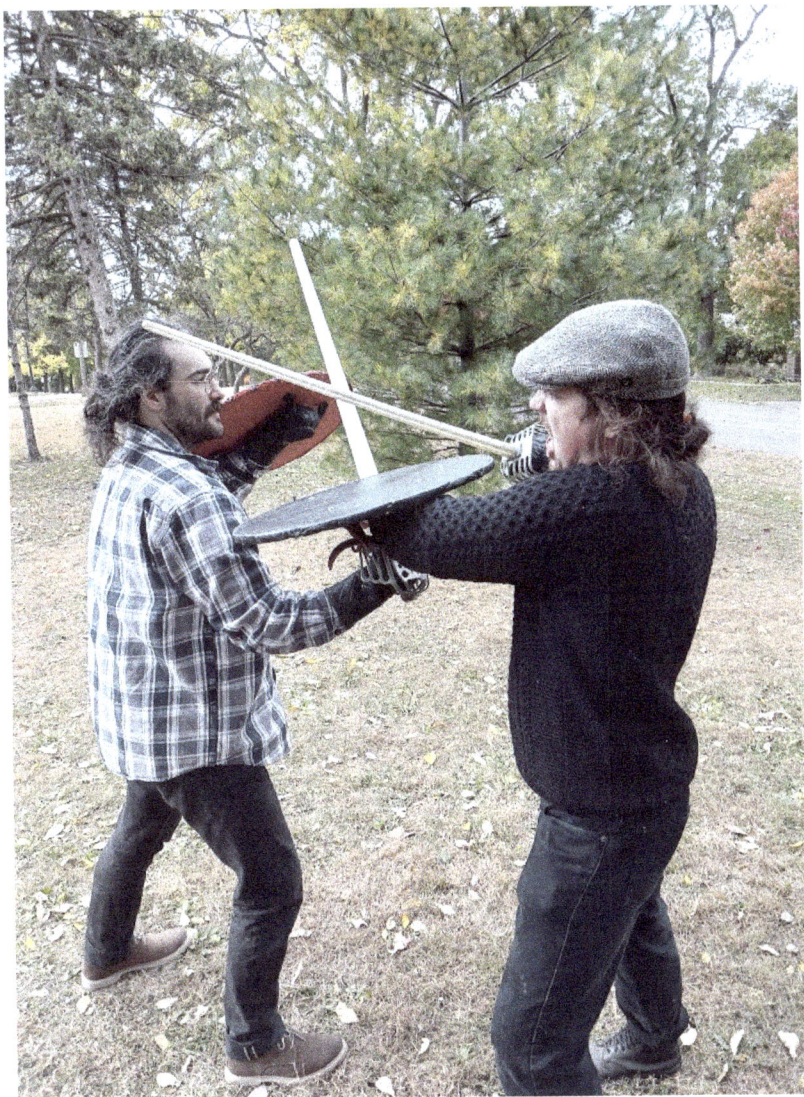

Suppresion

Targe Exercise

The Gladiator upon the Stage is very exact in these Lessons, and generally plays an exact round of them with little or no Variation: But the Highlanders in the Field make use of but a few of those Principles; but having another Instrument of defence turns his Sword chiefly to the Offensive Part, the outside and inside Throws are the Principle Offensive Uses of his Weapon; whilst he receives every Cut from his Adversary upon his Target which is a Shield fixt upon his Left Arm. (Page, 1746)

The targe can be used to parry any cut, but not by simply closing the line. If you try to parry with the targe in this static way, the opponent will simply redirect and hit you anyway. Instead, you should parry by punching your targe into the incoming cut.

The targe should be held at the angle that will most easily allow you to stop and control the incoming attack. For instance, a cut 1 should be parried while turning your targe hand as if making a jab, while a cut 2 should be parried with a lead hook.

The purpose of the targe exercise is to practice parrying with the targe and immediately cutting with your sword.

To practice the targe exercise, take an outside guard. Parry a high inside cut and return a high inside cut with your sword.

Take an outside guard. Parry a high outside cut and return a high outside cut with your sword.

Take an outside guard. Parry a low inside cut and return a high inside cut with your sword.

Take an outside guard. Parry a low outside cut and return a high outside cut with your sword.

Take an outside guard. Parry above your head with both weapons crossed and return a low inside cut with your sword.

Repeat the targe exercise at least ten times every time you practice the basic exercises.

Above: A jabbing parry

Below: A hooking parry

The Penicuik Guards

The standard sword and targe guards are the guards described by Thomas Page in 1746, but there is another (and quite possibly older) set of guards depicted in the Penicuik Sketches of Jacobite rebels, also from 1746.

The Open Guard: this is the position used in the Highland charge. To take this guard, step forward with your rear foot and raise your targe till it is next to your face, closing the line against any attack to your head on that side. For a right-handed fencer, this means that the left foot is forward rather than the right, and that the targe covers the left side of the head with the elbow tucked in. The sword is raised as if ready to chop right down.

The Underarm Guard: this is a strong counteroffensive position, and you can end up in this guard naturally after making a strong cut. This guard can also be seen in the Penicuik sketches. To take this guard, cut inside and let the sword arc around until it is pointed behind you, with the blade beneath your targe arm.

The Hanging Guard: can also be taken with a targe-foot-forward stance, as shown in the Penicuik Sketches.

The Low Guard: with the tip aimed at the ground is also a good counteroffensive stance. To take this guard, simply lower the sword from the open guard. This guard can be seen in the Penicuik sketches.

To practice the Penicuik Guards, simply move through the guards in the following sequence:

Open

Underarm

Hanging

Low

The open guard

Above: The Penicuik hanging guard
Below: The underarm guard

The Highland Charge Exercises

These exercises are based on the famous "Highland charge," the tactic of running straight at an enemy army with sword in hand. The Highland charge was essentially a psychological tactic, intended to panic the opposing forces into breaking and running – at which point the Highlanders could cut down their fleeing opponents from behind. The exercises are extremely simple for this reason.

The first five represent the charge itself, and the second five represent possible counters to a charge. The Highlanders used this tactic against enemy clans just as enthusiastically as they used it against government forces, so a Highland warrior would need to give some thought to how to resist an enemy charge. Unfortunately, any counter to the charge is only effective if your own side doesn't break and run, as it would be fatal to be the last person to start running!

Cut 2 and Cut 3

The first two Highland Charge Exercises are the most important, as they represent what a charging Highlander would do in the far majority of cases. They are based on the following passage from Thomas Page:

"The Highlander has nothing regular in Field Attacks and generally chop Right down to an Outside; or with a swinging and low Inside they endeavour to let out the Bowels, whilst every Part of his own Body is cover'd under a Target" (Page 1746).

"Chopping right down to an outside" is a cut 2, while a "swinging and low inside" is a cut 3.

The Bind

The third Highland Charge Exercise is based on the Bind, which is my name for the following technique described by Page:

"When two or three Throws have been made without Success, with your Body still square (that is your Legs crossing the Line of Defence at right Angles) and full facing your Adversary, drop both your Target and Sword as low as your Waste, your Sword still within your Target,and in that Posture lay your self open and wait for your Adversary's Throw, which when he makes, receive it not upon the Target, but upon the Fort of your Sword;

and at the same Moment by pushing your Target against his Hilt, drive his Sword sideways and downwards out of the Line, by which his Head will be expos'd defenceless; at which you may safely Throw, because his Sword will be held down by your Target, and his Left Arm and Target will be held down by his own Blade" (Page 1746).

In the third Charge Exercise this technique has been modified so you can perform it on a full run, but it appears in its original form in the sixth exercise. Two more variations of the Bind can be found in the seventh and eighth exercises.

The Lift

The fourth Charge Exercise is based on the Lift, which is my name for the following technique from Page:

"Another infallible Method both of Defence and Offence is, advancing briskly to your Adversary under an Inside Guard, receive his Outside upon your Fort, and at the same Moment instead of throwing an Inside, step briskly about with your Left Foot as in the Traverse (half a Circle at least) which will bring you under his Fort; and with your Target, which will be then under his Hilt, throw up his Sword and Arm, that you may have a free Passage for your own Sword, which you have lower'd and shortned in your coming about; and with a sudden Push slanting upwards, thrust in the Point between the Ribs on the Right Side, which commonly finishes the Affair" (Page 1746).

The fourth Charge Exercise is a variation on the Lift, which has been modified so you can perform it on the charge. The fifth and ninth exercises are also variations on the Lift.

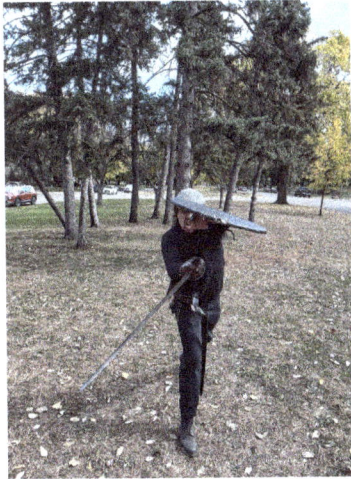

Above: (Left): Chopping right down to an outside with a cut 2
(Right): A swinging and low inside, or cut 3

Below: (Left) The Bind

(Right): The Lift

The Drop and Thrust

The Drop and Thrust is the last technique in the Highland Charge Exercises and is intended to counter the charge. It is essentially a combination of an image showing sword and shield combat on an early Scottish battle monument and a similar technique in McBane's smallsword manual.

The Guards

The charging exercises are performed from an open or high outside guard. This is essentially the same as Page's outside guard, except that the sword hand is raised high enough to allow you to make a cut 2 to the outside without having to raise your weapon first. If you do have to raise your sword to cut the outside with a cut 2, then you aren't holding your sword high enough. At the beginning of the charge, your targe foot should be forward in preparation for running. We'll refer to this position as the Penicuik stance, as this is the position shown in the Penicuik Sketches.

Exercise eight is performed from an underarm guard. This unusual guard is also found in the Penicuik Sketches. To take this guard, stand in the Penicuik stance, with the sword under the targe and pointed behind you, as if you have just completed a very wide cut.

Exercise ten is performed in the low guard. In the low guard, the sword is lowered and aimed at the ground.

Practicing the Highland Charge Exercises

The first five of these exercises are performed on a full run, with full-power cuts, and are therefore practiced solo as they would be too dangerous to practice with a training partner. It can be helpful to practice these exercises on a pell if you have access to one. The last five are practiced solo in a standing position.

In training, do not practice any of these sequences with a partner as it would be dangerous due to the speed and power of the attack when done on a full run. In a few of the photos, the exercise is shown partnered for clarity. The techniques in the Highland Charge exercise can be done with a partner only if done at controlled speed with no charge.

When charging, remember to cover with your targe by punching it out in the direction of the opponent's sword just before you make your cut. The idea here is to put yourself in a position to respond instantly to any attempted counter.

Feel free to begin your charge with a war cry if you prefer – the standard war cry for armies of mixed clans was "Claymore!"

1: Cut 2

Begin in the open or high outside guard. Charge at a full run until you reach your target, and make a chopping cut 2 at the target as you run by. When you make your cut, be sure to cover with your targe as a defense against a counterstrike.

2: Cut 3

Begin in the open or high outside guard. Charge at a full run until you reach your target, and make a slicing cut 3 at the target as you run by.

3: The Bind

Begin in the open or high outside guard. Charge at a full run until you reach your target, and make a chopping cut 1 or 2, followed instantly by a punch with the targe to knock the opponent's parry to the side. Circle your sword around over your head and make a cut 2, as if to the neck or the back of the head.

4: The Lift

Begin in the open or high outside guard. Charge at a full run until you reach your target, and make a chopping cut 2, followed instantly by a lift with the targe to knock the opponent's weapon up and then a stab to the body.

(Left): The Drop and Thrust. This should be practiced solo as it can be dangerous

(Right): Cut 2 on the charge

(Left): Cut 2 on the charge alternative side

(Right): Cut 3 on the charge

(Left): A bind on the charge

(Right): A lift on the charge

5: The Lift and Chop

Begin in the open or high outside guard. Charge at a full run until you reach your target, and make a chopping cut 2, followed instantly by a lift with the targe to knock the opponent's weapon up, then a stab to the body, and finally a cut to the head or neck.

6: Invite and Bind

Begin in an outside guard. Lower your weapons to your waist and await the charge. Parry the attack with your sword, punch the opponent's weapons aside with your targe, and circle the sword around to cut the head or neck.

7: Hanging Bind

Begin in a hanging guard. Parry the attack with your sword, punch the opponent's weapons aside with your targe, and circle the sword around to cut the head or neck.

8: Underarm Bind

Begin in an underarm guard. Parry the attack with your sword, punch the opponent's weapons aside with your targe, and circle the sword around to cut the head or neck.

9: Hanging Lift

Begin in a hanging guard. Parry the attack, lift your targe up to deflect the opponent's weapons, and stab the opponent.

10: Drop and Thrust

Begin in a low or medium guard. Raise your targe, drop to one knee, and stab up at the navel.

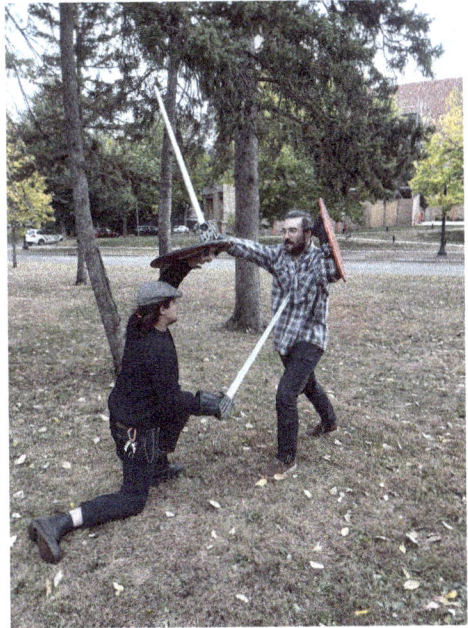

Above:(Left): Receiving the charge on the hanging guard

(Right): A drop and thrust to stop a charge.

Below: applying a bind from the hanging guard

The Ten Lessons of Sword and Targe

The Ten Lessons are choreographed fighting sequences, or "set play" in the terminology of the old broadsword manuals. Each lesson allows you to practice a technique or principle of sword and targe fencing in a controlled way. The lessons have been chosen from a mix of sources, including the fencing manuals of Thomas Page (1746), Donald McBane (1728), and Alexandre Valville (1817), as well as the Penicuik Sketches of Jacobite rebels from 1745-1746.

In the lessons, the person studying the technique is referred to as the protagonist, while the person facilitating the lesson is referred to as the antagonist. The antagonist's actions should be controlled and predictable so the protagonist can practice the technique. These lessons are a convenient way to learn and teach the techniques of sword and targe, but to apply them effectively in a bout you'll need to practice them against a resisting partner. You will have the opportunity to do this in the Bouting Exercises.

The Highland Charge Exercises, the Ten Lessons of Sword and Targe, and Bouting are the three required aspects of the Cateran Society Online Program's Sword and Targe curriculum.

The Guard of the Highlander

This lesson is derived from an illustration in Alexandre Valville's Treatise on the Contre-Pointe from 1817, a manual the French fencing master created for the Russian Imperial Guard. Valville includes a few illustrations purely for their interest, of which one is labeled "Guard of the Scottish Highlander" in both Russian and French. Valville shows a kilted man standing with a basket-hilted cudgel in a version of the prime hanging guard. His left arm is held up over his head, protected by what seems to be a rectangular home-made targe. This illustration appears to represent a Highland version of the traditional singlestick game based on the "broken head," with the addition of a targe.

To take the Guard of the Highlander, stand in the narrow stance with your targe over your head and your sword in a prime hanging guard.

The theme of this lesson is survival in melee combat. The sequence shown here is not strictly realistic but is intended to teach a simple strategy for survival in a melee situation.

In this position, you can "turtle up" under your weapons and absorb any number of attacks in a melee fight, lowering a weapon only when you need to in order to deal with a low attack. In a melee, most attacks will be simple downward cuts, from which you should be completely protected in this guard. At a moment of your choosing, you can parry on both your weapons simultaneously, then uncross your sword while leaving your targe on the opponent's weapon, cutting either high or low depending on the available opening.

Lesson One

Both fencers should take the Guard of the Highlander, a prime hanging guard with your targe held over your head.

Antagonist: lunge and cut the head.

Protagonist: parry with both weapons, then uncross and cut while keeping your targe on the opponent's sword. Cut the head if you can reach it but cut low inside if the line to the head is closed.

Note: don't push the targe over to the side but leave it horizontal over your head so that you can feel any movement in the opponent's sword through your targe. If you let your targe drift to the side, the opponent may be able to ride that movement downward and cut your leg.

The Guard of the Highlander

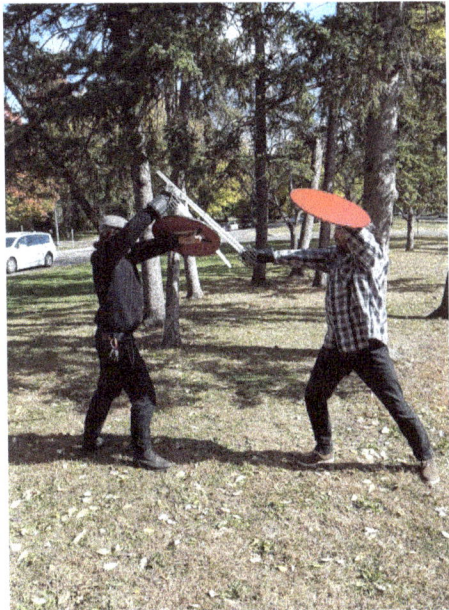
Receives the attack on both weapons

Riposte to the head

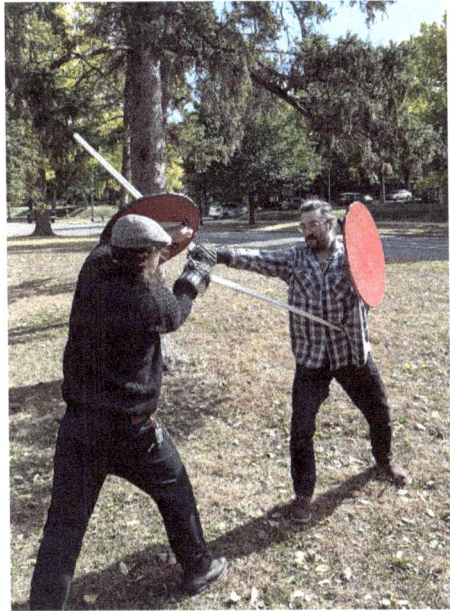
Alternative: riposte to the inside low

This lesson is based on two passages in the broadsword manuals: one from Donald McBane and the other from Thomas Page:

And if you cut at his leg, take care to cover your head with your buckler. (McBane, 1728)

With the target the cuts at the leg are differently made than without it, for under cover of that it is safe to go down to either outside or inside, without receiving a throw first. (Page, 1746)

It isn't safe to cut at the leg in a typical broadsword bout because the opponent will simply shift back and cut simultaneously at your arm or head. The situation is different with sword and targe because the targe can cover against an attack from above. Even if the opponent succeeds in slipping your attack, you should be able to prevent the opponent's counterattack from landing.

McBane describes this as covering the head, but you need to be aware of counters to your arm as well. The best approach is to punch out with your targe as you cut the leg, turning your hand so the targe forms a roof above you as you attack.

Lesson Two

Both fencers should take a hanging or outside guard.

Protagonist: lunge and cut the leg on the outside or inside, while turning your hand and punching out with your targe to cover against any attack from above.

Antagonist: stand still and receive the cut or shift back to slip the cut while cutting down at the head. Your sword will strike the protagonist's targe. Every once in a while – but not too frequently – counter to the arm rather than the head.

This lesson is derived from the following passage in Page:

When two or three throws have been made without success, with your body still square (that is your legs crossing the line of defence at right angles) and full facing your adversary, drop both your target and sword as low as your waste, your sword still within your target, and in that posture lay your self open and wait for your adversary's throw, which when he makes, receive it not upon the target, but upon the fort of your sword; and at the same moment by pushing your target against his hilt, drive his sword sideways and downwards out of the line, by which his head will be exposed defenceless; at which you may safely throw, because his sword will be held down by your target, and his left arm and target will be held down by his own blade. (Page, 1746)

Page's version of the bind begins from an invitation, a technique in which you leave yourself open to provoke an attack. In a real bout, the bind can be attempted after any parry with the sword. The bind is equivalent to a "command" or close-distance grappling technique with the broadsword and is similar (although not identical to) the "Turkish disarm."

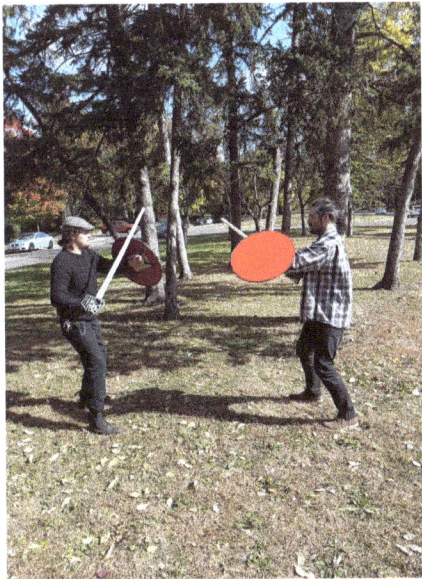

Start on the outside guard

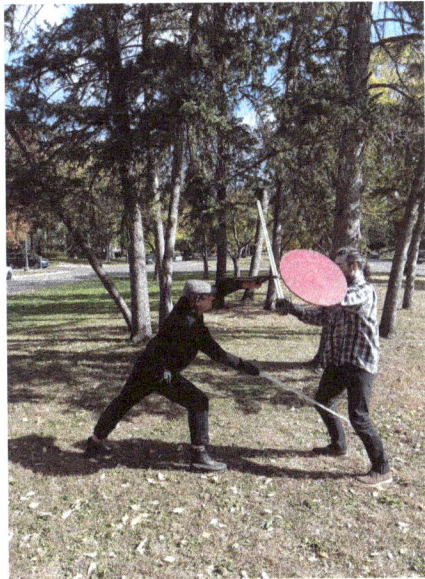

Cover the head when attacking the legs

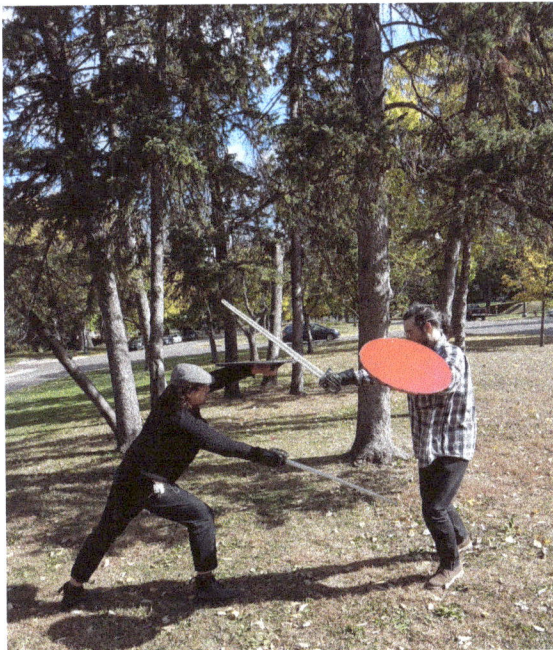

Negating the head counter attack

Lesson Three

Both fencers should take an outside guard.

Protagonist: lower your weapons to your waist to invite an attack.

Antagonist: step narrow and cut inside.

Protagonist: step narrow and parry with the sword on an inside guard, then pass forward with your rear foot and use your targe to bind the opponent's weapons over to your outside. Finish with a strong outside cut to the head or neck.

Note: be careful not to make contact with the opponent's neck, as it would be dangerous to strike this target even with a training sword.

Occasionally practice the bind in response to an outside cut rather than an inside cut. Other than the parry, the technique is exactly the same. Occasionally practice from a hanging guard without an invitation, by parrying on prime hanging and then binding.

The Lift

This lesson is based on the following passage:

Another infallible method both of defence and offence is, advancing briskly to your adversary under an inside guard, receive his outside upon your fort, and at the same moment instead of throwing an inside, step briskly about with your left foot as in the traverse (half a circle at least) which will bring you under his fort; and with your target, which will be then under his hilt, throw up his sword and arm, that you may have a free passage for your own sword, which you have lowered and shortened in your coming about; and with a sudden push slanting upwards, thrust in the point between the ribs on the right side, which commonly finishes the affair. (Page, 1746)

Again, Page shows this technique from an invitation, but it can also be performed without an invitation. The lift can be used after any parry with the sword on the outside guard.

Invite an attack

Parry on the inside

Bind and decapitate

Lesson Four

The antagonist begins in an outside guard, and the protagonist begins from out of distance on an inside guard, with the sword tucked up against the inside of the targe. The antagonist may find this lesson easier to perform from a "high outside" guard, allowing a strong cut to the outside with no need to chamber the cut.

Protagonist: take an inside guard and advance on the opponent as an invitation.

Antagonist: cut outside when the protagonist gets close enough.

Protagonist: parry outside with your sword, then lift the opponent's sword with your targe and thrust up between the ribs.

Occasionally practice from any outside parry with no invitation.

The Escape

This lesson combines Page's instructions for escaping the bind and the lift. In practice, the escape against the lift works equally well against the bind, so I teach this as the basic version of the escape, and I teach the escape against the bind as a more advanced variation.

Page's instructions for the escape are as follows:

These are the principle destructive methods of wounding in modern use; and when executed with a quick and a strong arm, and directed with a sharp and steady eye, seldom fail of success, except where an alert adversary is more steady at defence than your hand at throwing: in the last two cases indeed, no defence is practicable, if you suffer your self to be locked in the first, or to be closed upon the last; but how easy is the defence in either, when in the first, only by stepping into the back traverse, you at once free your sword, and by returning to your posture may wound your adversary, and be covered under your target; and in the last case, by retreating as he comes about with his left, you put your self out of the reach of his target, and much more out of that his sword, whilst he lies wholly exposed on his left side to your inside throw, how artfully soever, or how strongly soever it be made; but the same weapon which makes the attack, is capable of preventing the wound. (Page, 1746)

"Stepping into the back traverse" is the escape from the bind, and "retreating as he comes about with his left" is the escape from the lift. In this case, I interpret "retreating" as running backward rather than using a fencing retreat, because experiments have shown that it works much better.

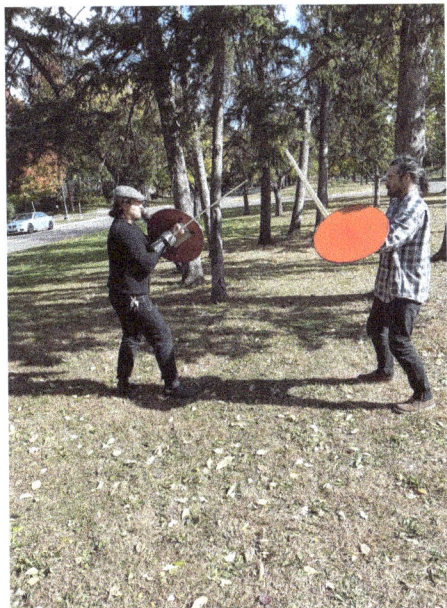

Advance behind an inside guard

Parry on the outside

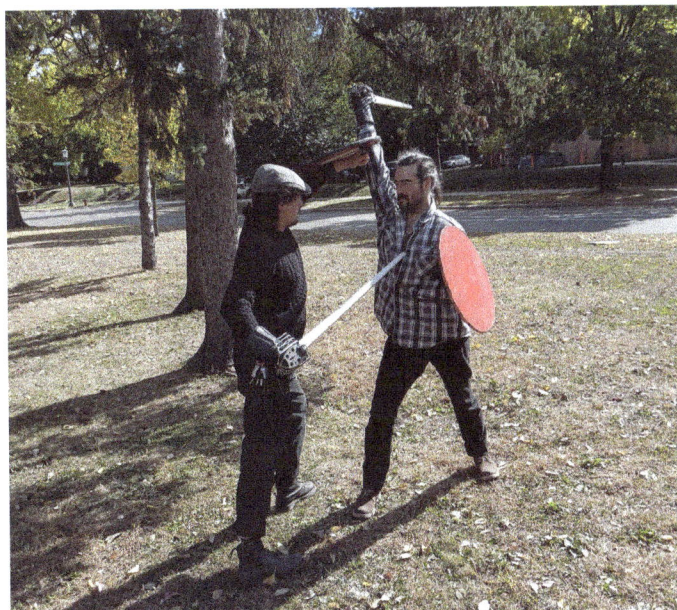

Lift and thrust

Lesson Five (Basic)

The antagonist takes an inside guard from out of distance, and the protagonist takes an outside or high outside guard.

Antagonist: advance on the protagonist in an inside guard. When they cut at your outside, attempt the lift.

Protagonist: cut outside when the antagonist gets close enough.

Antagonist: when the protagonist cuts at your outside, attempt the lift.

Protagonist: when the antagonist parries with the sword, run backward while simultaneously cutting inside low.

Lesson Five (Advanced)

Both fencers take the outside guard.

Antagonist: lower your weapons to your waist to invite an attack.

Protagonist: step narrow and cut inside.

Antagonist: step narrow and parry with the sword on an inside guard, then attempt the bind.

Protagonist: step into the back traverse to avoid the bind, then cut inside.

Note: the protagonist's step here is a simple back traverse; the foot should go to the side rather than backward.

Advance behind an inside guard

Parry on the outside

An attempted lift fails

Lower your weapon to invite an attack

Parry the inside cut

An attempted bind fails

An inside cut to the head ends the fight

Single Combat

This lesson is based on the following passage:

In single combat he aims at nothing more than disabling his antagonist which he commonly does by chopping him across the wrist within side the sword arm, which he does in the following manner; He runs up boldly to half sword, receives an outside, and changing with his adversary, drops his blade below the hilt upon the inside, draws the edge of his sword cross his adversary's wrist and springing backward saws it at the same time. (Page, 1746)

My interpretation of this technique has evolved over time, and my current interpretation incorporates the ideas of former Cateran Society instructor Jay Maas.

The technique was used to spare an opponent's life in single combat, a situation in which broadsword fencers typically tried to avoid the use of lethal attacks – unlike the battlefield, where the mentality was to "attack and not to be attackt" (Page, 1746).

Although the lesson as described by Page requires the antagonist to cut outside and then inside (or as later broadsword masters would have it, to cut 2 and then 1), the technique will actually work in modified form regardless of which combination the antagonist throws. The idea is simply to provoke a flurry of strikes by running in, then take the opportunity to cut the wrist.

Lesson Six

Both fencers take an outside or high outside guard.

Protagonist: run up to half-sword distance to provoke an attack.

Antagonist: as the protagonist approaches, cut outside and then inside.

Protagonist: parry the outside cut with your sword, then parry the inside cut with your targe while cutting the antagonist's wrist on the inside with a drawing cut as you spring back.

Parrying a cut to the outside on the charge

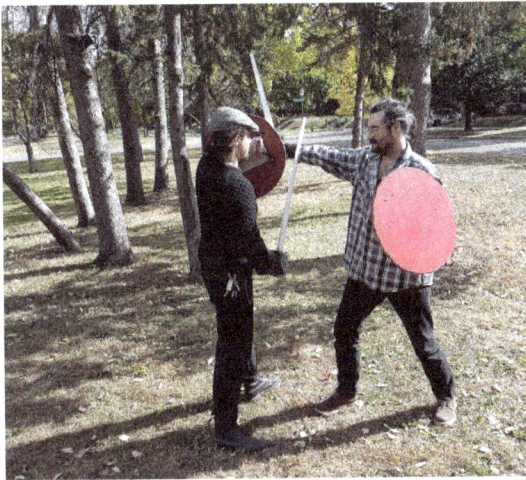

Cutting the wrist on the inside

Pitching to a Hanging

This lesson is based on the following passage:

Armed with a sword and a target being upon the left arm, advance to your enemy with a square body, and always under an outside guard, with your target advanced a little before your sword, and in a direction level with your adversary's breast, ready to receive any throw that he shall think fit to give; but wait not for it, it being safer to attack than be attacked, let your first throw be an inside betwixt your adversary's target and the sword; which if he receives upon the target, recover an outside, and pitch immediately to a hanging, but dwell not a moment upon it, but from that (which here is designed only to give a swing to your arm) throw home an inside at his left ribs underneath his left elbow, which will be opened by your pitching to a hanging, and by his raising a target to cover his head which will otherwise be exposed to be cut. (Page, 1746)

The act of "pitching" to a hanging guard functions like a feint, as the sudden upward movement will look much like a cut to a high target. In reality, you are simply chambering a low inside cut, while covering your own head against a counterattack by moving through the hanging guard. I teach a simplified form of pitching to a hanging as the basic version of this lesson, and the full sequence as described by Page as the advanced version.

Lesson Seven (Basic)

Both fencers take the outside guard.

Protagonist: pitch to a hanging, then make a low inside cut as one continuous movement.

Antagonist: respond to the feint by raising your weapons, giving the protagonist the opening to make the low inside cut.

Lesson Seven (Advanced)

Both fencers take the outside guard.

Protagonist: make an inside cut, then return to an outside guard.

Antagonist: parry the cut with your targe, then return to guard.

Protagonist: pitch to a hanging, then make a low inside cut as one continuous movement.

Antagonist: respond to the feint by raising your weapons, giving the protagonist the opening to make the low inside cut.

Note: the initial inside cut effectively plants the idea of a high attack in the antagonist's head, making them more likely to fall for the feint. This is how I interpret the line "let your first throw be an inside betwixt your adversary's target and the sword; which if he receives upon the target, recover an outside" (Page, 1746), although Paul Wagner of Stoccata School of Defence interprets this passage differently.

Pitch to a hanging

Cut to an inside low

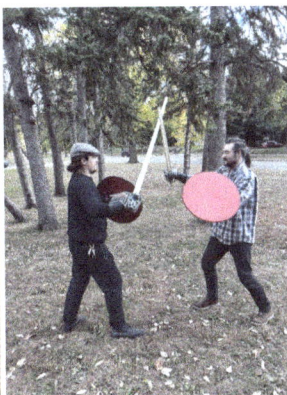
Start on an outside guard

Parry on an inside guard

Return to an outside guard

Pitch to a hanging guard

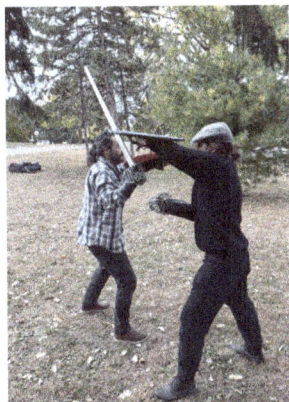
Cut to an inside low

This lesson is based on an illustration in Donald McBane's Expert Sword-Man's Companion of 1728, showing a man (possibly a soldier of the Edinburgh Town Guard) chopping with a Lochaber ax at a man with a sword and targe. The man with the sword and targe performs the technique described below to counter the cut.

Lesson Eight

Both fencers begin in an outside guard with sword and targe.

Antagonist: step in with an inside cut.

Protagonist: thrust into a hanging guard while stepping in, punching the targe hand straight forward so that the cut slides off the targe.

Note: the thrust must be performed with care to avoid injury; don't make contact with the tip. You will need fencing masks to practice this lesson in anything faster than slow motion. Occasionally experiment, by having the antagonist use a pole weapon such as the Lochaber ax rather than a sword and targe.

This lesson is based on the Highland Charge, as described by Page:

The Highlander has nothing regular in Field Attacks and generally chop Right down to an Outside; or with a swinging and low Inside they endeavour to let out the Bowels, whilst every Part of his own Body is cover'd under a Target. (Page, 1746)

This is the same technique found in the first two Highland Charge Exercises, except that it is not performed on a full run or with full power due to the presence of a training partner.

Lesson Nine

The protagonist should begin on the open or high outside guard. The antagonist should begin on a standard outside guard.

Protagonist: run in on the antagonist and chop outside or cut low inside as you go by. Just before you cut with your sword, punch your targe out at the opponent's basket-hilt while turning your targe so as to suppress any counterattack.

Note: the targe punch or "suppression" is an essential element of this technique. The idea is not to make hard contact with the antagonist's sword, but simply to be in position to immediately suppress any attempted counter. Turn the targe hand to a horizontal position as you punch.

Thrust with a hanging while defending with the targe

A chop outside on the charge

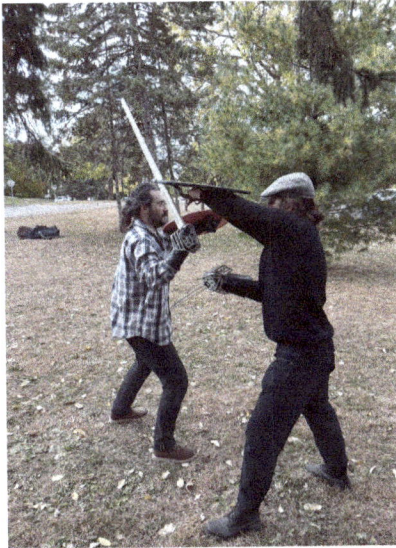

A swinging low outside cut on the charge

This lesson is based on the illustrations of the underarm guard in the Penicuik Sketches, and on Highland Charge Exercise number eight. Any other technique from the last five Highland Charge Exercises could be substituted for the technique shown here, and occasionally should be. As with lesson nine, be careful and do not use too much power when practicing with a training partner.

Lesson Ten

The antagonist should begin in an open or high outside guard. The protagonist should begin in an underarm guard.

Antagonist: run in on the antagonist and chop outside as you go by.

Protagonist: await the charge. Parry the attack with your sword, which will knock it far off-line. Bind with your targe while circling the sword around to the outside, then cut the head or neck.

Note: the parry and cut form one continuous, spiraling motion.

Wait in an underarm guard

Cut into the oncoming attack

Perform the bind

Remove the head

The Bouting Exercises

The set play lessons are really just an introduction to the core concepts of the sword and targe. To apply these concepts effectively in a real bout, you'll have to practice them in a noncooperative context. That's where the ten bouting exercises come in, as a series of fencing games with intentionally restricted rules designed to teach you how to apply the lessons.

These exercises are excellent for conditioning purposes, as they are all extremely exhausting if you do them seriously.

For all these exercises, you'll need the same protective gear you would need for an actual loose play bout: at minimum, a three-weapon mask with a back-of-the-head protector and a heavy jacket as well as a training broadsword and a targe.

Highland Singlestick Game

This exercise is based on the assumption that the "Guard of the Highlander" illustration represents a Highland version of the singlestick game, in which the only target was the head. The goal in the singlestick game was to "break the head" by inflicting a bleeding wound on it. Irish stickfighting instructor Maxime Chouinard has suggested that the only purpose of the strange arm targe in the "Guard of the Highlander" illustration may have been to serve as a target for head strikes in singlestick training.

While I prefer to interpret it as a crude homemade targe, I do agree that the illustration relates to singlestick training. The Memoirs of the Life and Gallant Exploits of the Old Highlander, speaking of Donald MacLeod, refers to singlestick play as the method by which Highland boys learned how to use the sword and targe: "his spare hours, like those of other boys, were wholly employed in training up himself, by cudgelplaying, to the use and management of the broad-sword and target" (Thomson 1791, 13). While MacLeod's Memoirs are not a wholly reliable source, this detail seems likely to be based on reality.

There are two main advantages to practicing this exercise. First, it will show you how to apply the seemingly very simple Lesson One in a real bout. Second, it will teach you one of the most important basic skills with sword and targe: the ability to cover up under both your weapons and weather any storm the opponent throws at you, picking your moment to come in with a decisive attack.

Rules: both fencers take the Guard of the Highlander with singlesticks or other broadsword trainers as well as targes. When the game begins, both fencers may attempt to hit each other on the fencing mask only, keeping their targes high to make the mask as hard as possible to hit. Any other strikes are to be ignored. The action should be continuous, with no pauses in between touches. It continues until one fencer steps back out of distance and lowers their weapons. As an advanced exercise, you can allow strikes to other targets for the purpose of forcing the opponent to lower their defense or suffer the pain of being struck – but touches to the mask are still the only touches that count.

The Leg Attack Game

The purpose of the second game is to teach you how to keep your head safe when you attack the leg, as instructed by both Donald McBane and Thomas Page. The game will also teach you another important piece of information: as you will see for yourself, the head cannot be reached without exposing your leg to a potential cut.

Rules: both fencers take the Guard of the Highlander. When the game begins, both fencers may attempt to hit each other on either the fencing mask or the leg, taking care to cover the head whenever they attack the leg. Any touches other than touches to the mask or leg are to be ignored. You can also play this game from any other guard.

The Binding Game

The third game teaches you how to apply the bind in a real bout. As you will soon discover, it is much harder to achieve a bind in a fencing match than in the controlled conditions of a set play lesson! For this game to work properly, the antagonist must attack continuously and without pause until a touch is made by either fencer (unlike in the first two games, you can pause and reset after a touch).

Rules: designate one fencer as the protagonist and the other as the antagonist. Both fencers take the outside guard (the protagonist can use the invitation if they want to, but it is not required). When the game begins, the antagonist launches a series of attacks without pausing, while the protagonist attempts to parry with the sword, apply the bind with the targe, and finish the bout. When the protagonist achieves a successful bind, the roles switch for the next turn. As an advanced version, you can play this game from any other guard.

The Lifting Game

This game teaches you to apply the lift in a real bout – which thankfully turns out to be much easier than applying the bind. As in the previous game, the antagonist must attack continuously and without pause until a touch is made. Note that the lift can generally only be done after an outside parry with the sword.

Rules: designate one fencer as the protagonist and the other as the antagonist. Both fencers take any guard (the protagonist can invite by fencing from the inside guard if they want to, but it is not required). When the game begins, the antagonist launches a series of attacks without pausing, while the protagonist attempts to parry outside with the sword, apply the lift with the targe, and finish the bout. When the protagonist achieves a successful lift, the roles switch for the next turn.

The Escaping Game

Page teaches two different techniques for escaping a command with the targe: one for the bind, and one for the lift. I teach the lift escape as the basic version because it will also allow you to escape from the bind, but the bind escape is still an option. This game teaches you to escape either the bind or the lift, and it is quite effective – if you go into the game with the intention of using the lift escape at the first opportunity, it should work almost every time. (The bind escape is harder to pull off, although not impossible.) Note that the protagonist, not the antagonist, is the one attacking here.

Rules: designate one fencer as the protagonist and the other as the antagonist. Both fencers take any guard. When the game begins, the protagonist launches a series of attacks without pausing. The antagonist attempts to achieve either a bind or a lift. The protagonist attempts to apply either version of the escape and to strike the antagonist in doing so.

The Single Combat Game

You might have noticed that the Single Combat lesson is a little artificial: after all, who's to say that the antagonist will respond with a cut two-cut one combination attack? As you will see from this game, the specific example given by Page is just that

– an example. The lesson will work against any combination attack the antagonist throws, as long as you apply the concept flexibly and opportunistically. Therefore, the act of running in on the opponent functions like an invitation: the idea is to provoke them to launch an aggressive series of attacks to drive you back, giving you the opportunity to time their sword arm.

Rules: designate one fencer as the protagonist and the other as the antagonist. Both fencers take an outside guard. When the game begins, the protagonist runs in, and the antagonist launches any combination of attacks as the protagonist approaches. The protagonist tries to parry with the targe, then cut the antagonist's sword arm as it comes in for another attack. (In some cases, the targe parry is preceded by a sword parry, while in other cases it is not. That depends on the combination the antagonist uses.)

The Pitching Game

This game will teach you how to apply the concept of "pitching to a hanging" from lesson seven. The basic version of this lesson is simply to swing your sword up from an outside guard to a hanging guard to cause your opponent to twitch upward a little in anticipation of a high attack. This gives you an opening, while also chambering a cut 3 to the opponent's torso. The advanced version of the lesson is the version described by Page, which begins with a cut to the inside to put the idea of a high attack in your opponent's head before pitching to a hanging.

Rules: designate one fencer as the protagonist and the other as the antagonist. Both fencers take an outside guard. When the game begins, the protagonist makes a variety of attacks to a mix of high and low targets, while the antagonist merely defends without attacking. At any point, the protagonist may pitch to a hanging and cut under with a cut 3.

The Hanging Thrust Game

This game teaches you how to use McBane's hanging thrust. While McBane's technique is shown in use against the Lochaber ax, it will also work against the sword and targe. Note that it is generally not possible to bout safely with the Lochaber ax due to its weight and power.

Remember that the targe must be punched straight forward, without turning the fist, while the sword makes the hanging thrust. The opponent's attack should slide harmlessly down to the side.

Rules: designate one fencer as the protagonist and the other as the antagonist. Both fencers take an outside guard. When the game begins, the antagonist launches a series of attacks without pausing, and the protagonist responds with the hanging thrust. Use this technique with care and don't let the intensity level get too high, as a thrust with the singlestick can be dangerous.

The Highland Charge Game

This game teaches you how to use the Highland charge in a real bout. Unlike the Highland Charge Exercises, you should not run at full speed or make your attacks with full power, as that would be dangerous. For the protagonist, this exercise will be bouting exercise number nine. For the antagonist, it will be number ten.

Rules: designate one fencer as the protagonist and the other as the antagonist. The protagonist takes a high outside guard, and the antagonist takes any guard. The protagonist runs in, suppressing with the targe and cutting either 2 or 3. The antagonist attempts to counter the charge using any of the last five techniques from the Highland Charge Exercises.

The Countering Game

This game teaches you how to counter the Highland charge. For the protagonist, this exercise will be bouting exercise number ten. For the antagonist, it will be number nine. Take extra care to avoid injury.

Rules: designate one fencer as the protagonist and the other as the antagonist. The antagonist takes a high outside guard, and the protagonist takes any guard. The antagonist runs in, suppressing with the targe and cutting either 2 or 3. The protagonist attempts to counter the charge using any of the last five techniques from the Highland Charge Exercises.

The Lessons of Thomas Page

The Ten Lessons of Sword and Targe include several techniques from Thomas Page, interspersed with techniques from Donald McBane, Alexandre Valville, and the Penicuik Sketches. By contrast, this set of lessons is drawn exclusively from Page's 1746 work The Use of the Broadsword. You'll find several of them familiar from your previous work with the Ten Lessons, but you'll also find some new, and in some cases, challenging techniques. These are given in the same order in which they appear in Page's work. In the following lessons, P stands for the protagonist and A stands for the antagonist.

1: The Slip

In the Field of Battle and in promiscuous Combat his first Principle is to attack and not to be attackt, and his Attack begins at all Times with a full Throw at the outside of the Sword Arm; which if he misses, instead of changing to an Inside, he makes a push at the Navel with the Point of his Sword, but not going home, is ready to slip his Adversary, who will infallibly throw at that wide Opening he has given to his Head and upper Part of his Body; and if he succeeds in the Slip, with a full Lunge he throws an Outside to his Adversary's Neck, which for the most Part severs the Head from the Body (Page, 1746)

The slip in this lesson would be Page's "inside slip" : When your Adversary Throws an Inside, instead of Stopping it with an Inside Guard, draw your Right Foot backward towards the Left, in the same Manner as in the Retreat, and at the same Moment withdraw your whole Body backward and Sideways to the right of the Line, letting your Adversary's Point pass by your Sword a little out of his Reach, and steping into your former Position, Throw home at his Outside, which can't but be open by his over throwing himself, which He will do the more by missing your Body, and not being receiv'd by your Sword, which he expected, to stop the effort of his Strength. This is the Slip upon the Inside.

In other words, your lead foot should shift back to rear foot while you pull your sword in on an inside guard, and when the opponent misses you should lunge forward and cut the outside.

On the outside guard. P cuts outside, A slips the cut. P feints a thrust at the navel, A cuts at the head inside. P slips the cut, then cuts the neck or head on the outside.

2: The Feint

In the Field of Battle and in promiscuous Combat his first Principle is to attack and not to be attackt, and his Attack begins at all Times with a full Throw at the outside of the Sword Arm; which if he misses, instead of changing to an Inside, he makes a push at the Navel with the Point of his Sword, but not going home, is ready to slip his Adversary, who will infallibly throw at that wide Opening he has given to his Head and upper Part of his Body; and if he succeeds in the Slip, with a full Lunge he throws an Outside to his Adversary's Neck, which for the most Part severs the Head from the Body: But if his Adversary makes no Attempt to throw at the Opening, he returns to his push in reality and stabs him a little above the Navel; which will oblige his Adversary to lower his Sword and give him that Opening at his Head and Neck which he in vain attempted before, and which he will now be sure to hit and for the most Part split the Scull. (Page, 1746)

Instructions

On the outside guard. P cuts outside, A slips the cut. P feints a thrust at the navel, A doesn't respond. P thrusts to the body, A attempts to parry by sweeping downward with both weapons. P avoids the parry, disengages, and cuts down at the head.

3: Field Attacks

The Highlander has nothing regular in Field Attacks and generally chop Right down to an Outside; or with a swinging and low Inside they endeavour to let out the Bowels, whilst every Part of his own Body is cover'd under a Target. (Page, 1746)

Instructions

On the outside guard, P runs at A and cuts outside, or cuts low inside to "let out the bowels."

4: Disabling

In single Combat he aims at nothing more than disabling his Antagonist which he commonly does by chopping him across the Wrest within Side the Sword Arm, which he does in the following Manner; HE runs up boldly to half Sword, receives an Outside, and changing with his Adversary, drops his Blade below the Hilt upon the inside, draws the Edge of his Sword cross his Adversary's Wrest and springing backward saws it at the same Time. (Page, 1746)

Instructions

On the outside guard, P runs at A. A makes a double attack, outside and inside. P parries the outside and times the inside, sawing the wrist. P springs off.

5: Pitching to a Hanging

Arm'd with a Sword and a Target being upon the Left Arm, advance to your Enemy with a square Body, and always under an Outside Guard, with your Target advanc'd a little before your Sword, and in a Direction levell with your Adversary's Breast, ready to receive any Throw that he shall think fit to give; but wait not for it, it being safer to attack than be attacked, let your first Throw be an Inside betwixt your Adversary's Target and the Sword; which if he receives upon the Target, recover an Outside, and pitch immediately to a Hanging, but dwell not a Moment upon it, but from that (which here is design'd only to give a Swing to your Arm) throw home an Inside at his Left Ribs underneath his Left Elbow, which will be open'd by your pitching to a Hanging, and by his raising a Target to cover his Head which will otherwise be expos'd to be cut. (Page, 1746)

Instructions

On the outside guard, P cuts inside. A parries with the targe. P returns to the outside guard, then pitches to a hanging. A raises the targe to parry, and P cuts under at the ribs with a low inside.

6: The Leg Attack

With the Target the cuts at the Leg are differently made than without it, for under Cover of that it is safe to go down to either Outside or Inside, without receiving a Throw first. (Page, 1746)

Instructions

On the outside guard. P cuts the leg while covering the head with the targe.

7: The Bind

When two or three Throws have been made without Success, with your Body still square (that is your Legs crossing the Line of Defence at right Angles) and full facing your Adversary, drop both your Target and Sword as low as your Waste, your Sword still within your Target,and in that Posture lay your self open and wait for your Adversary's Throw, which when he makes, receive it not upon the Target, but upon the Fort of your Sword; and at the same Moment by pushing your Target against his Hilt, drive his Sword sideways and downwards out of the Line, by which his Head will be expos'd defenceless; at which you may safely Throw, because his Sword will be held down by your Target, and his Left Arm and Target will be held down by his own Blade. (Page, 1746)

Instructions

On the outside guard, P lowers both weapons. A attacks, P parries with the sword. P passes forward with the rear foot, binds A's weapons over with the targe, and finishes with the sword.

8: The Lift

Another infallible Method both of Defence and Offence is, advancing briskly to your Adversary under an Inside Guard, receive his Outside upon your Fort, and at the same Moment instead of throwing an Inside, step briskly about with your Left Foot as in the Traverse (half a Circle at least) which will bring you under his Fort; and with your Target, which will be then under his Hilt, throw up his Sword and Arm, that you may have a free Passage for your own Sword, which you have lower'd and shortned in your coming about; and with a sudden Push slanting upwards, thrust in the Point between the Ribs on the Right Side, which commonly finishes the Affair. (Page, 1746)

Instructions

A on the outside guard, P on the inside guard. P advances, and A cuts outside. P parries outside with the sword, passes forward with the rear foot, uses the targe to lift A's sword up, and thrusts up into the body.

9: Escaping the Bind

These are the Principle destructive Methods of Wounding in Modern Use; and when executed with a quick and a strong Arm, and directed with a sharp and steady Eye, seldom fail of Success, except where an alert Adversary is more steady at Defence than your Hand at Throwing: In the last two Cases indeed, no Defence is practicable, if you suffer your self to be lock'd in the first, or to be clos'd upon the last; but how easy is the Defence in either, when in the first, only by stepping into the Back Traverse, you at once free your Sword, and by returning to your Posture may wound your Adversary, and be cover'd under your Target; (Page, 1746)

Instructions

On the outside guard, A lowers both weapons. P attacks, A parries with the sword. P steps into the back traverse to escape the bind, then counters.

10: Escaping the Lift

and in the last Case, by retreating as he comes about with his Left, you put your self out of the Reach of his Target, and much more out of that his Sword, whilst he lies wholly expos'd on his Left Side to your Inside Throw, how artfully soever, or how strongly soever it be made; but the same Weapon which makes the Attack, is capable of preventing the Wound. (Page, 1746)

Instructions

P on the outside guard, A on the inside guard. A advances, and P cuts outside. A parries outside with the sword and attempts to apply the lift. P runs backward to escape, while cutting inside low.

The Penicuik Lessons

These lessons are an advanced version of the Ten Lessons of Sword and Targe, using the same techniques but relying exclusively on the guards shown in the Penicuik Sketches. By practicing this version of the lessons, you should be able to see how any of the sword and targe techniques could have been used with the Penicuik guards. Instructions are given for the protagonist only, as the antagonist's actions should be obvious once you have reached the stage of practicing the Penicuik Lessons.

1: The Hanging Guard

Take a hanging guard. Parry inside with both weapons and cut inside while keeping your targe on the antagonist's sword.

2: Covering the Head When Attacking the Legs

Take an open guard. Cover your head with your targe while cutting the legs.

3: The Bind

Take an open guard. Lower your weapons to your waist to invite an attack. Pass forward with your rear foot and parry with your sword, then pass forward again and use your targe to bind the opponent's weapons over to your outside. Finish with a strong outside cut.

4: The Lift

Take a low guard. Parry a chop with your targe, then use the targe to lift the opponent's sword up while thrusting at the navel or between the ribs.

5: The Escape

Take any guard and make a cut, but spring off with a counter when your opponent parries with the sword to initiate a bind or lift. You can escape either with a back traverse followed by a cut inside, or by simply running backward and cutting inside low.

6: Single Combat

Take an open guard. Run up to half-sword distance, parry an outside cut with the sword, then time the opponent's inside cut by cutting the wrist on the inside.

7: Pitching to a Hanging

Begin on an open guard. Make an inside cut, then return to an open guard and pitch to a hanging before cutting inside low.

8: The Hanging Thrust

Begin on an open guard. Make a hanging thrust into the opponent's attack to stop it. The antagonist can use a pole weapon such as the Lochaber ax, or a sword and targe.

9: The Highland Charge

Begin on the open guard. Run in on the opponent and chop outside or cut low inside as you go by. Don't forget to punch out with your targe just before your cut.

10: Countering the Charge

Take an underarm guard and await the charge. Parry the attack with your sword, bind with your targe, and cut the head or neck.

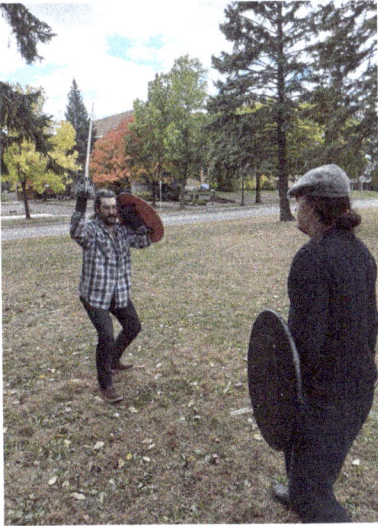

Charging on an open guard into an invitation

Cutting into the attack to the outside

Performing the bind

Loose Play

"Loose play" was the old term for free fencing with the broadsword, while "set play" referred to choreographed drills. You can practice loose play according to any ruleset you prefer, but you should always wear enough protective gear to allow you to bout safely. This includes a three-weapon fencing mask, a back-of-head protector, and any other gear you may need in order to bout without injury.

Protecting the head while attacking the legs

Suppressing an attack with the targe to cut to the inside

Parrying inside creating an oppertunity to bind

Crowding with the targe to close in

Cutting to the outside

Thrusting under the targe

Parrying outside to create an opening for a lift

Advanced Sword and Targe Lessons

These advanced sword and targe lessons give you the opportunity to expand your sword and targe skills by examining how the lessons in the other broadsword manuals would change if a targe was added to the mix. The idea in every case is to try out the lesson several times, working out variations based on the presence of the targe.

This is a good opportunity to experiment and discuss your results with your training partner. In some cases, the sequence will have to be modified due to the presence of either your targe or your training partner's targe. In other cases, you will find techniques that don't make any sense to attempt if a targe is in play.

As you practice, ask yourself the following questions:

1. How would I change this attack due to the presence of a targe?

2. Can I use the targe to prevent this attack?

3. What tactical lessons can I draw from this?

The ten lessons here are the same as those taught in Level I of the Cateran Society's broadsword system: Plain Playing, Timing, Slipping the Leg, Slipping the Body, Double Attacks, Feints, Invitations, Actions on the Blade, Commands, and Counter-Disarms.

Every example given is based on a sequence from one of the historical broadsword manuals, and the manual is indicated in parentheses after the technique. However, in some cases the sequence is changed a little from its original form. Use the circular, traversing footwork with every lesson, then experiment with other types of footwork if you wish. Parries should generally be done with the targe.

Once you've studied all the variations of all ten lessons, you should have a solid understanding of what can be done with a sword and targe and how the targe would affect a broadsword bout.

Plain Playing

On the hanging guard, P cuts 1 and A parries. A cuts 1 and P parries. P cuts 1 and A parries. A cuts 4 and P parries. (McBane 1728)

On the hanging guard, P cuts 1 and A parries. A cuts 1 and P parries. P cuts 1 and A parries. A cuts 1 and P parries. P cuts 3 at the ribs. (McBane 1728)

On the outside guard, A cuts 1. P steps in toward the cut and parries with the targe, then cuts the leg. (Page, 1746)

On the outside guard, A cuts 1. P steps forward and parries inside, then performs a draw cut 2 at the face inside the opponent's guard and flies off. (Ferdinand, 1788)

On the inside guard, A cuts 2. P steps forward and parries outside, then performs a draw cut 1 at the face inside the opponent's guard and flies off. (Ferdinand, 1788)

Timing

Timing an Outside: On the outside guard, A changes to inside guard and P times the wrist. (Page, 1746)

Timing an Inside: On the inside guard, A changes to outside guard and P times the wrist. (Page, 1746)

Timing a Hanging: On the outside guard, A changes to hanging guard and P times the arm. (Page, 1746)

The Lurch: On the hanging guard. A feints 1 and cuts 2. P darts the tip of the sword at A's face to forestall the final cut. (Lonnergan, 1771)

Both fencers traverse. A raises the sword to cut, and P times the arm with cut 1 or 3, followed by a cut 2 while traversing to the opponent's outside. (Ferdinand, 1788)

Slipping the Leg

The Leg Attack: On the outside guard. P cuts 4 at the leg while covering the head with the targe. A shifts and counters to the head, and P parries with targe. (McBane, 1728)

The Leg Attack: On the outside guard. P cuts at the leg while covering the head with the targe. A shifts and counters to arm rather than head, and P parries with targe. Watch the timing here – it can be difficult for P to avoid being struck in the arm, despite the presence of the targe. (McBane, 1728)

Slipping the Body

The Inside Slip: On the outside guard, A cuts 1, P performs the inside slip by stepping back till the heal of the lead foot touches the rear foot, while pulling back into an inside guard. P then cuts 2. Note how the targe can be used to suppress any follow-up attack. (Page, 1746)

The Outside Slip: On the inside guard, A cuts outside, P slips the cut by stepping back until the lead foot moves behind the read foot, crossing the line. At the same time, P takes an outside guard, then cuts 1. (Page, 1746)

The Hanging Slip: On the hanging guard, A cuts 1, P performs the hanging slip by lunging diagonally forward and right, then cuts 1. Note how the targe provides extra protection during the hanging slip. (Page, 1746)

Double Attacks

On the hanging guard. A cuts 1 and P parries. P cuts 1 and A parries. P traverses and cuts 4. (McBane 1728)

On the outside guard. P cuts 3 at the wrist then 4 at the wrist. (Lonnergan, 1771)

On the outside guard. P cuts 3 at the wrist then 2 at the face. (Lonnergan, 1771)

On the outside guard. P cuts 3 at the wrist then 2 at the face, then 3 at the wrist springing back. (Lonnergan, 1771)

On the inside guard. P cuts 4 at the wrist then 3 at the wrist. (Lonnergan, 1771)

On the inside guard. P cuts 4 at the wrist then 1 at the face. (Lonnergan, 1771)

On the inside guard. P cuts 2 at the face then 3 at the wrist springing back. (Lonnergan, 1771)

On the outside guard. P cuts 3 at the wrist then 4 at the wrist, then 3 at the wrist springing back. (Lonnergan, 1771)

On the inside guard. P cuts 4 at the wrist then 3 at the wrist, then 4 at the wrist springing back. (Lonnergan, 1771)

On the inside guard. P cuts 4 at the wrist then 1 at the face, then 4 at the wrist springing back. (Lonnergan, 1771)

On the inside guard. P cuts 4 at the wrist then 1 at the face then 2 at the face, then 3 at the wrist springing back. (Lonnergan, 1771)

The Feint

On the outside guard. P feints 1, then cuts 2. (McBane 1728)

On the inside guard. P feints 2, then thrusts inside. (McBane 1728)

On the hanging guard, P feints 1 and cuts 1. (Page, 1746)

On the hanging guard, P feints 7 and cuts 1. (Page, 1746)

On the hanging guard, P feints 1 and cuts 2. (Page, 1746)

On the outside guard, P cuts 3 at the leg. A slips and cuts 1, P parries while remaining on the lunge. P cuts 4 at the leg and hits. (Page, 1746)

On the outside guard, P drops on deeply bent knees and feints at the leg. A slips and cuts 1, P parries, cuts at the leg and hits. (Page, 1746)

On an outside guard. P feints 3 at the wrist, then cuts 4 at the wrist. (Lonnergan, 1771)

On an outside guard. P feints 1 at the face, then cuts 4 at the wrist. (Lonnergan, 1771)

On an outside guard. P feints 1 at the face, feints 2 at the face, then cuts 3 at the wrist or body. (Lonnergan, 1771)

On an outside guard. P feints 3 at the wrist, feints 4 at the wrist, then cuts 3 at the wrist. (Lonnergan, 1771)

On an outside guard. P feints 1 at the face, feints 2 at the face, then cuts 1 at the head. (Lonnergan, 1771)

On an inside guard. P feints 2 at the arm, then cuts 1 at the arm. (Lonnergan, 1771)

On an inside guard. P feints 4 at the body, then cuts 3 at the body. (Lonnergan, 1771)

On an inside guard. P feints 2 at the face, then cuts 3 at the wrist in drawing back. (Lonnergan, 1771)

On an inside guard, P feints 4 at the thigh, then cuts 3 at the thigh. (Lonnergan, 1771)

On an inside guard, P feints 4 at the leg, then cuts 3 at the wrist. (Lonnergan, 1771)

On an inside guard. P feints 2 at the face, feints 1 at the face, then cuts 2 at the face. (Lonnergan, 1771)

On an inside guard. P feints 2 at the face, feints 1 at the face, then cuts 4 at the wrist. (Lonnergan, 1771)

On an inside guard. P feints 2 at the face, feints 1 at the face, then cuts 4 at the body. (Lonnergan, 1771)

On an inside guard. P feints 2 at the face, feints 1 at the head, then cuts 2 at the arm. (Lonnergan, 1771)

On a hanging guard. P feints 1 at the face, then cuts 1 at the head. (Lonnergan, 1771)

On a hanging guard, P feints 1 at the head, then cuts 3 at the wrist. (Lonnergan, 1771)

On a hanging guard, P feints 1 at the head, then cuts 3 at the body. (Lonnergan, 1771)

On a hanging guard, P feints 1 at the head, then cuts 4 at the body. (Lonnergan, 1771)

On a hanging guard. P feints 1 at the head, feints 4 at the body, then cuts 1 at the head. (Lonnergan, 1771)

On a hanging guard. P feints 1 at the head, feints 2 at the arm, then darts the point underneath. (Lonnergan, 1771)

On the outside guard, P feints at A's leg. A slips and counters to head. P parries with targe, then cuts A's leg outside. (Ferdinand, 1788)

The Invitation

On the inside guard, P lowers both weapons to invite an attack. A cuts 1, and P performs the inside slip and cuts 2. (Page, 1746)

On the outside guard, P advances and then lowers both weapons to leave an opening on the inside. P appels by stomping the foot while vocalizing Ha! repeatedly. A cuts 1, and P times the arm. (Page, 1746)

P lowers both weapons to invite. A cuts 1. P crouches a little and parries with targe, then cuts 4 at the leg. (Lonnergan, 1771)

P lowers both weapons on the outside guard. A cuts 2, and P times the cut with a cut 2. (Ferdinand, 1788)

P lowers the tip on an inside guard. A cuts at the arm outside, and P makes a simultaneous parry outside and cut outside to the arm or face. (Ferdinand, 1788)

Actions on the Blade

Bearing: On the outside guard, P lunges and bears (pushing the opponent's blade down), then cuts inside the guard. (Page, 1746)

Battering: On the outside guard, P batters (a violent beat) and cuts 2. (Page, 1746)

The Ambuscade: On the outside guard, P advances to half-sword and bears. A disengages and cuts inside, and P times the wrist with 3. (Page, 1746)

On an inside guard. A makes a double attack of 2 and 1. P parries the first cut outside and the second cut inside while stepping in. P whirls to the hanging guard to disarm A or open the line, then finishes. (Lonnergan, 1771)

Commands

On the hanging guard. A thrusts, P parries with outside half-hanger and performs the bind. (Sinclair, 1790)

On the outside guard. P bears on A's blade, runs up hilt hilt, and applies the lift. (Mathewson, 1805)

Counter-Disarms

On the outside guard. A bears on P's blade and attempts a lift. P shifts back and thrusts to the body. (Mathewson, 1805)

A bears on P's blade and attempts a lift. P shifts back and cuts 4. (Mathewson, 1805)

Sword, Targe and Dirk Lessons

Highlanders sometimes fought with the sword, targe, and dirk in which case the dirk was held in the same hand that held the targe. These lessons allow you to explore this interesting combination.

Lesson 1: Take an outside guard. The opponent cuts inside. You parry with your sword in an inside guard while closing in rapidly. Hook your dirk over their sword near the hilt, pin it down and finish them with your sword.

Lesson 2: Take a hanging guard. The opponent attacks inside low. Parry with the portion of the dirk that protrudes below the targe as you close in and cut 3.

Lesson 3: Take an outside guard. The opponent makes cut 4. Parry outside low, pass forward with your left foot, and stab the opponent in the head with the dirk.

Lesson 4: Take an outside guard. The opponent makes cut 1 then cut 2. Parry cut 1 with your targe. When the opponent makes cut 2, parry on an outside guard, pass forward with your left foot, and stab them in the head with the dirk.

Lesson 5: You make cut 1. The opponent parries with their sword and attempts to apply the Bind. Pass forward with your left foot and dirk them.

Lesson 6: The opponent charges at you while making cut 1. Parry with your targe, and push forward so that your dirk threatens their face as you thrust to their body with your sword. Take great care when practicing this technique and go slow at first.

Lesson 7: The opponent cuts at the head. Parry St. George (a horizontal head parry) with your sword as you run in, using your sword to Lift the opponent's sword up. Slice their wrist with your dirk.

Lesson 8: Make cut 1, then cut 2. The opponent parries both attacks on the sword and attempts to apply a Lift. As they do so, you suddenly pass forward with your left foot and dirk them.

Lesson 9: The opponent cuts at the head. Parry St. George with your sword as you run in, using your sword to Lift their sword up. Stab them in the face with your dirk.

Lesson 10: When the opponent cuts at the head, parry on St. George with your sword and then run in, using the targe and dirk to apply a Lift from below. Finish them with a simultaneous dirk to the face and sword to the body.

Certification in Sword and Targe

The Cateran Society offers certification in sword and targe fencing in Level Three of its curriculum. To earn certification in sword and targe, complete the first two levels of the Cateran System (Regimental Highland Broadsword and Old-Style Broadsword) with a certified Mentor or Cateran or through the Cateran Society Online Program. Then, study Level Three. Requirements for certification include:

1: Practice the Ten Highland Charge Exercises to your Mentor's satisfaction.

2: Memorize the Ten Lessons of Sword and Targe or any of the other sets of ten lessons given in this book.

3: Fight a successful Certification Bout against a representative of another school of swordsmanship.

Once your Mentor certifies you, you will be a Level Three Mentor, certified to give instruction in sword and targe fencing.

If you do not choose to study sword and targe fencing through the Cateran Society, you can still learn the art from this book with a training partner.

I recommend training in this order:

1: Practice the Basic Exercises. If you have completed the first two levels of the Cateran System, you don't need to spend a lot of time on these. If you haven't, you should not start bouting until you've done these exercises for about six months.

2: Once you have some experience with the Basic Exercises, add in the Highland Charge Exercises.

3: Once you have some experience with the Highland Charge Exercises, add in the Ten Lessons of Sword and Targe.

4: Once you have experience with the Ten Lessons, add in the Bouting Exercises.

5: Once you have some experience with the Bouting Exercises, begin Loose Play. Bouting Exercises and Loose Play should represent the majority of your Sword and Targe training once you are ready for them.

6: If you like, try out the Lessons of Thomas Page. These are optional.

7: If you like, try the Penicuik Lessons. These are also optional.

8: Whenever you feel ready, experiment with the Advanced Lessons. These are meant as experiments, so practice them at your own pace. These are also optional.

9: If you like, try out the ten Sword, Targe, and Dirk lessons. These are optional as well.

10: If you like, incorporate Sword, Targe, and Dirk into Loose Play.

To learn Sword and Targe fencing, the only exercises you absolutely need are:

The Basic Exercises

The Highland Charge Exercises

The Ten Lessons of Sword and Targe

Loose Play

However, the Bouting Exercises will make it far easier for you to build real skill in Loose Play.

Enjoy fencing with the Sword and Targe!

www.ingramcontent.com/pod-product-compliance
Lightning Source LLC
Chambersburg PA
CBHW070937280326
41934CB00009B/1910